Customary and Metric Units
(4-6)

© 2015 OnBoard Academics, Inc
Portsmouth, NH
800-596-3175
www.onboardacademics.com
ISBN: 978-1-63096-084-1

OnBoard Academic's books are specifically designed to be used as printed workbooks or as on-screen instruction. Each page offers focused exercises and students quickly master topics with enough proficiency to move on to the next level.

OnBoard Academic's lessons are used in over 25,000 classrooms to rave reviews. Our lessons are aligned to the most recent governmental standards and are updated from time to time as standards change. Correlation documents are located on our website. Our lessons are created, edited and evaluated by educators to ensure top quality and real life success.

Interactive lessons for digital whiteboards, mobile devices, and PCs are available at www.onboardacademics.com. These interactive lessons make great additions to our books.

You can always reach us at customerservice@onboardacademics.com.

Units of Measure

Key Vocabulary

metric measure

customary measure

length

mass

weight

Can you order these customary and metric units of measure?

Customary Units			Metric Units	
Weight	**Length**		**Mass**	**Length**
		GREATEST		
		LEAST		

Millimeter Pound   Kilogram   Meter   Ton   Centimeter

Kilometer   Foot Yard   Ton   Ounce   Gram   Mile Inch

Draw a line to connect the equal customary units of measure.

100 yd	= 3 ft =	2 yd
1 mi	= 72 in =	36 in
6 ft	= 1,760 yd =	3,600 in
1 yd	= 300 ft =	5,280 ft

Measuring Lengths: Customary units of measure.

User your ruler to measure these items.

_____ in

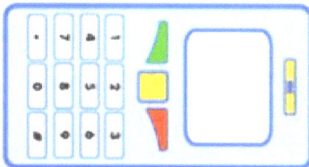

_____ in

_____ in

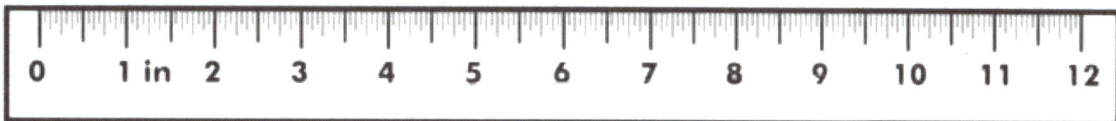

Can you match each image with an appropriate metric measurement.
Write you answer in the blue box.

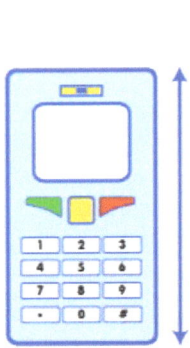

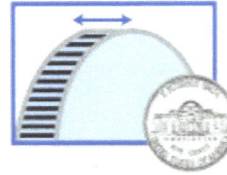

| 12 mm | 9.3 cm | 177 m | 1,300 km |
| 28 cm | 2 mm | 130 km | 1.77 m |

Measure the lengths of these objects to the nearest Millimeter.

| | cm = | | mm |

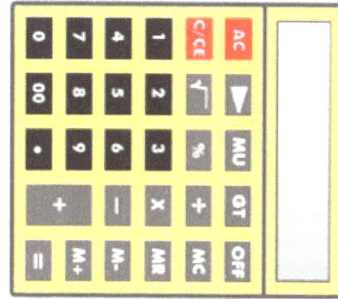

| | cm = | | mm |

| | cm = | | mm |

Customary units of weight.

Match the item with its approximate weight.

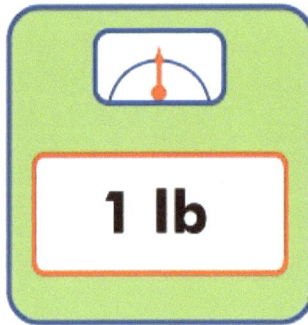

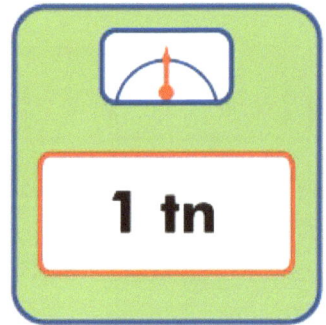

What is the most appropriate unit of measure for the mass of each object below?

56 kg   0.4 t   20 g   2 g   5.6 kg   4 t

Approximate customary to metric measure conversions

1 in ≈ 2.5 cm

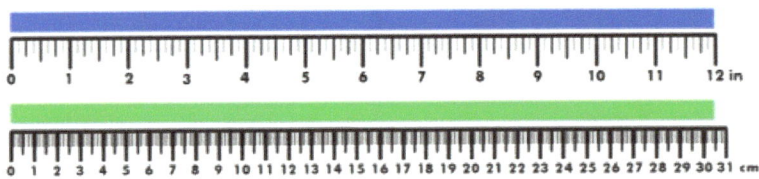

1 ft ≈ 0.3 m

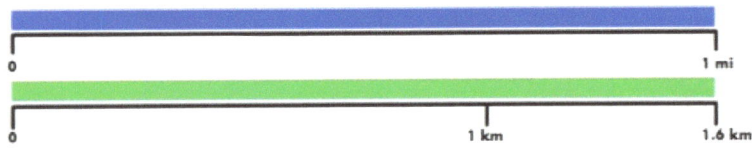

1 mi ≈ 1.6 km

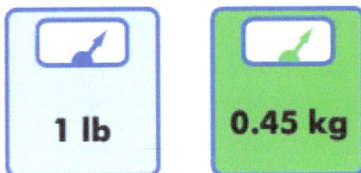

1 lb ≈ 0.45 kg

Practice converting between customary and metric measurement units.

5 in ≈ ⬜ cm?

3 ft ≈ ⬜ m?

8 mi ≈ ⬜ km?

$7\frac{1}{2}$ lb ≈ ⬜ kg?

Name: _____

Units of Measure Quiz

1 True or false, a mile is shorter than a kilometer?

2 Which of the following statements is *not* correct?

A   10 ft = 120 in

B   17,600 ft = 10 mi

C   72 in = 2 yd

D   5 yd = 15 ft

3 5.5 m = ? cm?

4 12 in ≈ ? cm?

30	20	10	48
A	B	C	D

Customary Measure

Key Vocabulary

customary units

weights

capacity

Match the customary units of measure.

100 yd		2 yd
	= 3 ft =	
1 mi	= 72 in =	36 in
	= 1,760 yd =	
6 ft		3,600 in
	= 300 ft =	
1 yd		5,280 ft

Measure the length of each item.

_____ in

_____ in

_____ in

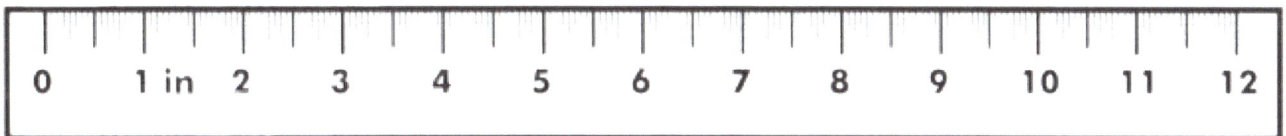

Weight

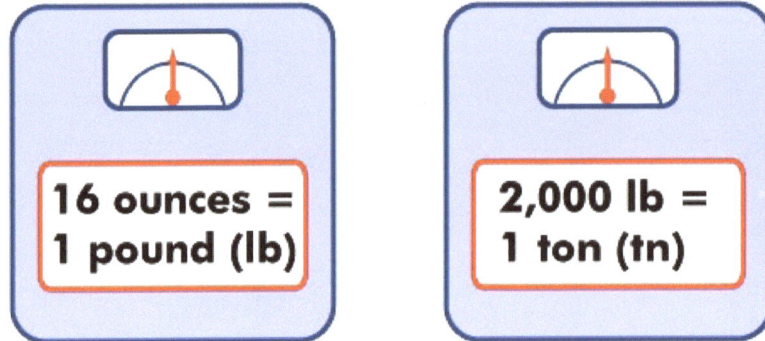

Under each scale list 3-5 things that weigh approximately the amount shown on the scale.

A few items are listed to get you started

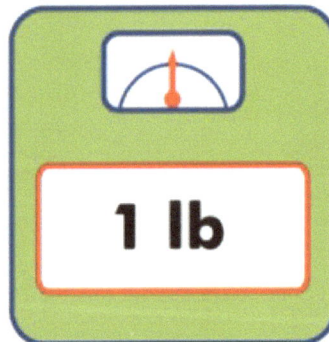

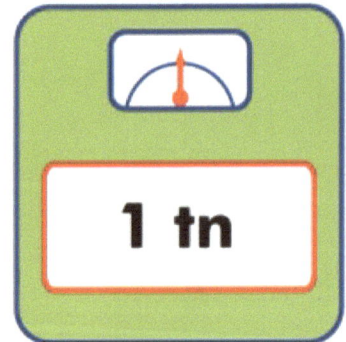

Toast   Loaf of Bread   Small Car

How heavy is each object?
Circle the weight you estimate.

2 oz	**5 oz**	**4 lb**
6 oz	**2 lb**	**12 oz**
3.5 lb	**5 lb**	**2 oz**
8 lb	**8 lb**	**1 lb**

Capacity

1 gallon (gal) = 4 quarts (qu)
1 quart = 2 pints (pt)
1 pint = 2 cups (c)
1 cup = 8 ounces (oz)

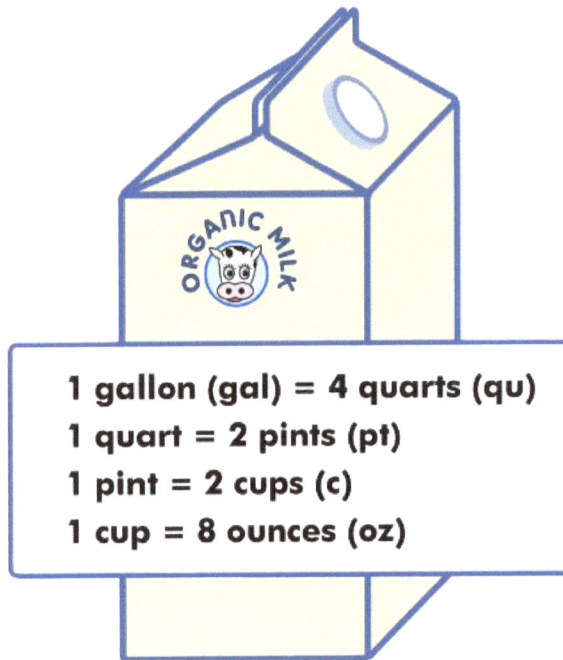

Match the can segments.

8 pt　　**0.5 gal**　　**4 qu**　　**2 qu**　　**2 gal**

4 pt　　**4 pt**　　**8 qu**　　**16 cups**　　**1 gal**

Name_____

Customary Measure Quiz

1 True or false? 1 mile = 1,762 yd

2 What is the length of this pencil?

(A) 9 in

(B) $9\frac{1}{4}$ in

(C) $9\frac{3}{8}$ in

(D) $9\frac{1}{2}$ in

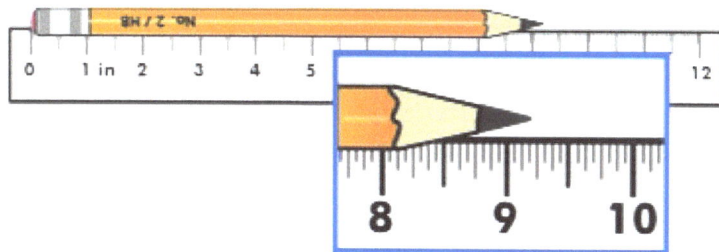

3 $3\frac{1}{2}$ lb = ___ oz?

4 5 gallons = ___ pints?

Metric Measure for Mass & Capacity

Key Vocabulary

metric units

mass

capacity

Measuring length: mm and cm.

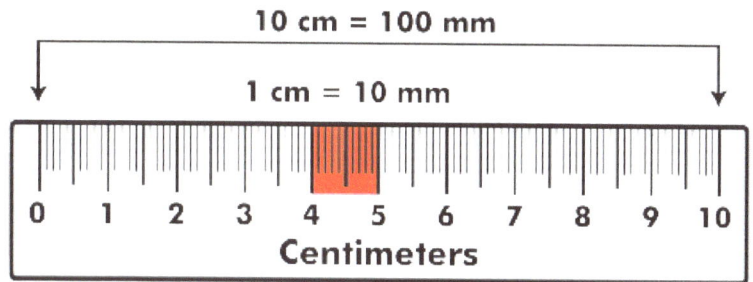

10 cm = 100 mm

1 cm = 10 mm

0 1 2 3 4 5 6 7 8 9 10
Centimeters

3 cm = [ ] mm

7.3 cm = [ ] mm

111 mm = [ ] cm

Measuring length: m, cm and mm.

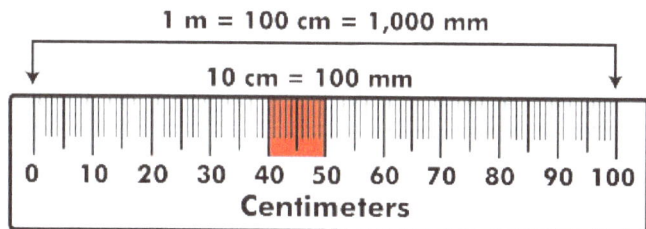

1 m = 100 cm = 1,000 mm

10 cm = 100 mm

0 10 20 30 40 50 60 70 80 90 100
Centimeters

0.5 m = [ ] cm

0.5 m = [ ] mm

150 cm = [ ] m

Measuring length: km, m, cm and mm.

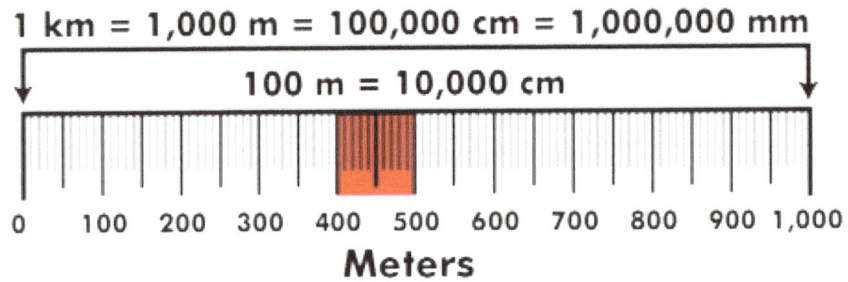

1 km = 1,000 m = 100,000 cm = 1,000,000 mm

100 m = 10,000 cm

0   100   200   300   400   500   600   700   800   900 1,000

Meters

1.5 km = [   ] m

1.5 km = [   ] cm

600 m = [   ] km

Write the appropriate metric measurement from the suggestions given.
Measure each items.

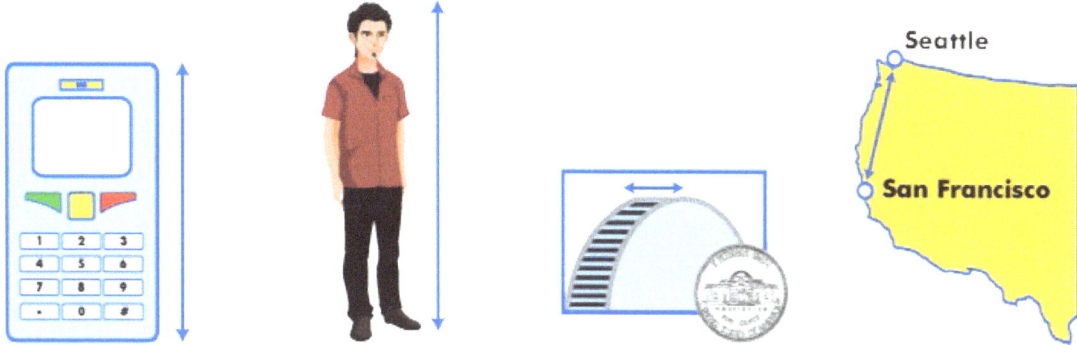

12 mm	9.3 cm	177 m	1,300 km
28 cm	2 mm	130 km	1.77 m

[ ] cm = [ ] mm

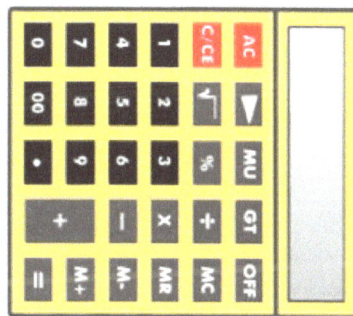

[ ] cm = [ ] mm   [ ] cm = [ ] mm

Convert metric units of mass.

1 g = 1,000 mg

1 kg = 1,000 g

1 t = 1,000 kg

1.5 kg = [ ] g

7.5 g = [ ] mg

0.65 t = [ ] kg

1,500 mg = [ ] g

3 t = [ ] g

What is the most appropriate mass for each item..
Choose from the suggestions given.

56 kg   0.4 t   20 g

2 g   5.6 kg   4 t

Estimate the capacity of each of these containers in ml or l.

Name_____

Metric Measure for Mass and Capacity Quiz

1 True or false, metric measures for capacity include liters, centimeters and milliliters?

2 Which of the following statements is *not correct*?

A 1 m = 1,000 mm

B 1 l = 1,000 ml

C 1 km = 1,000 cm

D 0.01 km = 1,000 cm

3 Write 5.5 liters in milliliters.

4 Write 12,000 m in km.

www.ingramcontent.com/pod-product-compliance
Lightning Source LLC
Chambersburg PA
CBHW052045190326
41520CB00002BA/200